ıe Quantum Threshold

ere quantum computing actually stands, what one more year buys, and what it will — and will not — tell us about universe.

tian Magnuson
: Advisory Group
: edition · August 2026

out this volume

is a research briefing for people who have to make decisions, not a survey of the literature. It ers the twelve months ending August 2026, projects one year forward, and separates what has ı demonstrated from what has been announced.

so takes seriously a set of questions that most business coverage of quantum computing ds and most popular coverage abuses: what quantum computers will and will not tell us about strong nuclear force, dark matter, entanglement, the early universe, wormholes and extra ensions. Chapter 8 answers those directly, including where the honest answer is no.

Evidence discipline. Every statistic in this volume is attributed. Where a figure originates with a vendor rather than an independent party, the text says so. Company roadmaps are described as targets, never as forecasts. Where a capability is emerging rather than deployed, it is labelled emerging — and where the market overclaims, naming the overclaim is treated as part of the analysis rather than a digression from it.

w to read it in ten minutes

d the Executive Summary, then Chapter 4 for the one-year projection, then Section 9.1 for the work in this book that cannot wait. If your organization simulates molecules or materials, add tions 5.2 and 5.3. If you want the physics, Chapter 8 stands alone.

urcing

ıary sources — peer-reviewed journals, arXiv preprints, company technical disclosures and stor filings, and government documents — supplemented by a small number of specialist ets with editorial track records. Appendix D lists them. Appendix B gives the five-question filter d to screen every claim that appears here.

CONTENTS

xecutive Summary

antum computing crossed a real line in the twelve months ending August 2026, and it is important to be precise about which line rossed.

id not cross into commercial usefulness. It did not break encryption. It did not simulate a drug molecule that classical chemistry could not dle. What it crossed is narrower and, for anyone making decisions on a five-year horizon, more consequential: the field moved from *iming* it could outperform classical computers to *proving* it, on problems that independent parties can check, using verification protocols igned to survive hostile scrutiny.

at is the story of the past year. Google's Quantum Echoes result in October 2025 produced the first quantum advantage claim that was erministic and repeatable rather than statistical and contestable. Nine months later, in July 2026, IBM published three separate advantage ims in four days — one with a certified fidelity bound, one cross-validated on a competitor's trapped-ion hardware, one with an open-rced classical simulator so critics could try to knock it down. Quantinuum's Helios demonstrated 48 logical qubits from 98 physical ones, a o that would have looked like a typo three years ago. QuEra reported 96 logical qubits in January 2026. Zuchongzhi 3.2 in China went below fault-tolerance threshold with a distance-seven surface code.

ne of these machines can do your job. All of them are evidence that the engineering path to machines that can is now visible rather than oretical.

second story of the year is that the security clock moved faster than the capability clock. Three papers between May 2025 and March 2026 the estimated cost of breaking RSA-2048 and elliptic-curve cryptography by roughly two orders of magnitude — not through better dware, but through better algorithms. Craig Gidney's revision took RSA-2048 from twenty million physical qubits to fewer than one million. ogle's March 2026 elliptic-curve paper put 256-bit ECC at fewer than 500,000 qubits with a runtime in minutes. Scott Aaronson — for two ades the field's most reliable brake on overclaiming — wrote in April 2026 that hardware experts he trusts now think a cryptographically vant quantum computer is possible around 2029, and framed the post explicitly as a warning.

third story is the widest gap: between what quantum computing will plausibly do for physics and what popular coverage claims it will do. antum computers are already producing publishable results in lattice gauge theory, cosmological particle creation, and quantum-chaos amics. Quantum sensors — a different technology that shares the same physics — are the leading instruments in the search for axion and k-photon dark matter. Those are real. Quantum computers will not build a wormhole, will not enable time travel, and did not "create" one in 2 despite headlines saying so. The distinction between quantum computing as a *simulation instrument for theories we already have* and ntum computing as a *magic key to the universe* is the single most useful thing a reader can take from this book.

at to expect by August 2027, on the evidence in this volume: several hundred to low thousands of physical qubits per module across ding platforms, with tens to low hundreds of logical qubits demonstrated; IBM's Kookaburra and Cockatoo modules proving qLDPC mory and inter-module entanglement; Quantinuum's Sol shipping; a small number of quantum-advantage results that survive a year of ssical counterattack; continued absence of commercial quantum advantage in optimization, machine learning, or drug discovery; and PQC gration moving from voluntary to mandatory across regulated sectors under the June 2026 executive orders.

operating question for a business leader is not whether quantum computing works. It is which of your assets have a cryptographic shelf life ger than five years, and who owns the answer.

Chapter 1 — What a Quantum Computer Actually Is, and What "Power" Means

1.1 The one-paragraph version

A classical bit is definitively zero or one. A qubit occupies a *superposition* — a weighted combination of both, described by amplitudes that can be positive, negative, or complex. Link many qubits through *entanglement* and the system's state is described by amplitudes across all possible configurations at once: *n* qubits require 2^n complex numbers to specify. Fifty qubits already exceed what a laptop can hold; three hundred exceed the number of atoms in the observable universe. A quantum algorithm arranges interference between those amplitudes so that wrong answers cancel and right answers reinforce. The computation is not "trying all answers in parallel." It is sculpting a wave so that the answer is what's left standing.

This distinction matters commercially. Quantum computers are extraordinary at problems whose structure is itself quantum — molecules, materials, gauge fields — and at a small number of structured mathematical problems where the interference pattern happens to be constructible, factoring being the famous one. They are ordinary or worse at most everything else, and the "everything else" category includes the majority of business optimization problems that vendors have historically pitched.

1.2 Why counting qubits tells you almost nothing

Through roughly 2023, the industry's headline metric was raw physical qubit count. That metric is now close to useless, and the industry knows it.

Physical qubits are noisy. Every gate operation has an error probability. Run a circuit of a few thousand gates on a device with a 0.1% two-qubit error rate and the signal is gone. What matters is the *product* of qubit count, gate fidelity, connectivity, and circuit depth — how many useful operations you can chain before noise wins.

The field has converged on three better framings:

Logical qubits. Quantum error correction encodes one protected "logical" qubit across many physical qubits, using redundancy and repeated measurement to detect and correct errors without collapsing the state. Below a certain physical error rate — the *fault-tolerance threshold* — adding more physical qubits per logical qubit makes the logical qubit exponentially better rather than worse. Google's Willow demonstrated this crossover in December 2024. Everything since has been engineering it into systems.

Error-corrected operations. Riverlane's proposed "QuOps" metric counts error-free quantum operations, with a generational ladder — KiloQuOp, MegaQuOp, GigaQuOp, TeraQuOp — deliberately echoing how mobile networks are versioned. IBM frames the same idea as gate counts on logical qubits: its 2029 Starling target is one hundred million gates on two hundred logical qubits.

Verified advantage on a defined task. The most honest metric, and the one 2026 was actually about: can you name a specific computation, run it, and have the result survive a year of people trying to reproduce it classically?

The physical-to-logical ratio is the number to watch. In November 2024, Atom Computing and Microsoft produced 24 logical qubits from 1,180 physical qubits — a 49:1 overhead. In November 2025, Quantinuum's Helios produced 48 logical qubits from 98 physical qubits, a 2:1 ratio, using a new code family. In January 2026, QuEra reported 96 logical qubits from 448 physical qubits, roughly 4.7:1. Those ratios are not directly comparable — they encode different code distances, protection levels, and operational capabilities — but the trend is unmistakable and it is the trend that decides whether fault tolerance arrives in 2029 or 2039.

1.3 The four hardware platforms, and why the field has not picked one

Superconducting circuits (Google, IBM, Rigetti, IQM, Alice & Bob, Chinese academic programs) use microwave-frequency circuits on chips cooled to about 15 millikelvin. Gates are fast — tens of nanoseconds — and the manufacturing borrows directly from semiconductor fabrication, which is why IBM's shift to 300mm wafers matters. The weaknesses are limited connectivity (qubits mostly talk to neighbours), the sheer physical footprint of dilution refrigerators, and coherence times measured in microseconds.

Trapped ions (Quantinuum, IonQ, Oxford Ionics before its acquisition) hold individual charged atoms in electromagnetic traps and manipulate them with lasers. Fidelities are the best in the industry — Oxford Ionics reported 99.99% two-qubit gate fidelity, and Quantinuum's Helios averages 99.921%. All-to-all connectivity comes free because ions can be shuttled. The cost is speed: gates are microseconds to milliseconds, thousands of times slower than superconducting.

Neutral atoms (QuEra, Pasqal, Atom Computing, Infleqtion, and as of March 2026 a new Google lab in Boulder) hold neutral atoms in optical tweezer arrays. This platform scales in *space* better than anything else: Caltech trapped 6,100 caesium atoms in September 2025, with 13-second coherence and 99.98% single-qubit control fidelity, and Harvard demonstrated a 3,000-qubit system running continuously — reloading atoms without shutting down, which had been a hard practical ceiling. Gates are slower than superconducting and the platform is younger.

Photonics (PsiQuantum, Xanadu, Quandela, ORCA) computes with light. Photons barely decohere and can run at room temperature outside the chip, and the manufacturing leverages existing photonic foundries. The difficulty is that photons do not interact with each other, so gates must be built from measurement and heralding, with brutal loss budgets.

Topological qubits (Microsoft) attempt to build error protection into the physics rather than layering it on in software. Microsoft's Majorana 2, reported June 2026, achieved parity lifetimes around 22 seconds and roughly doubled the topological gap to about 70 microelectronvolts by switching from aluminium to lead and redesigning with indium arsenide. The scientific community's response has been careful: the 2025 Majorana 1 claims drew sustained criticism, and the 2026 paper describes a prototype multi-tetron structure, not a processor. This is the highest-variance bet in the industry — if it works, it shortcuts the overhead problem entirely; if the underlying quasiparticles are not what Microsoft believes, the programme has no floor.

Google's March 2026 move is the most revealing datapoint about platform convergence. Rather than doubling down on superconducting, Google opened a neutral-atom lab in Boulder under JILA's Adam Kaufman. Hartmut Neven's stated logic: superconducting scales in the time dimension — circuit depth — while neutral atoms scale in the space dimension — qubit count. When the company that achieved the first below-threshold error correction hedges across modalities, that is a statement that nobody has won.

1.4 The classical half of a quantum computer

The least discussed development of the past year is how much *classical* compute a fault-tolerant quantum computer needs. Error correction requires decoding syndrome measurements in real time — reading which errors occurred and computing the correction faster than new errors accumulate. At scale this is a substantial streaming-inference problem with a hard microsecond-scale latency budget.

NVIDIA's NVQLink, announced October 2025, is the industry's answer: an open architecture tightly coupling GPUs to quantum processors specifically for calibration, error decoding, and control. It launched with 17 QPU builders — Alice & Bob, Atom Computing, Diraq, Infleqtion, IonQ, IQM, ORCA, Oxford Quantum Circuits, Pasqal, Quandela, Quantinuum, Quantum Circuits, Quantum Machines, Quantum Motion, QuEra, Rigetti, SEEQC, Silicon Quantum Computing — and nine US national laboratories including Brookhaven, Fermilab, Berkeley, Los Alamos, Oak Ridge, PNNL and Sandia.

The practical implication: a quantum computer is a peripheral inside a classical supercomputer, not a replacement for one. Quantinuum's deployment of Helios inside Oracle Cloud Infrastructure AI data centres, and its HPE integration partnership, follow the same logic. Anyone modelling quantum as a standalone category is modelling it wrong.

Chapter 2 — The Year in Review: August 2025 to August 2026

2.1 The verification turn

The single most important intellectual shift of the past year is that "quantum advantage" stopped being a marketing word and started being a claim with a burden of proof.

Google's 2019 supremacy result used random circuit sampling — a task with no use, and a result that could only be verified statistically. Classical algorithms subsequently narrowed the gap, and the episode taught the field an expensive lesson about claims that cannot be checked.

Quantum Echoes (Google, 22 October 2025). Running on Willow's 105-qubit processor using a 65-qubit subsystem, Google implemented an algorithm based on out-of-time-order correlators — measuring how a perturbation propagates and "echoes back" through a chaotic quantum system. Google reported roughly 13,000× faster execution than the best classical approach it tested, with individual data points estimated at three-plus years each on Frontier versus seconds on Willow. Critically, the output is deterministic and repeatable, so an independent party can rerun it and cross-check. In collaboration with UC Berkeley, the team applied the method as a "molecular ruler," reconstructing geometry for two molecules of 15 and 28 atoms via simulated NMR spin-echo signals — results that matched conventional NMR while surfacing additional structural detail. This is a proof of concept, not a chemistry product. It is also the first time a quantum advantage claim came attached to a physically meaningful measurement.

IBM's July 2026 triple. In four days, three claims landed on IBM Heron hardware.

- IBM with the University of Chicago (28 July) demonstrated advantage via "doped Clifford sampling" — circuits mostly built from efficiently simulable Clifford gates with strategically placed non-Clifford T gates, which makes the output checkable. On 70 data qubits plus 27 syndrome ancillas, they certified a state-fidelity lower bound of 0.284 at 95% confidence.
- Qedma (27 July) ran Floquet Ising magnetization dynamics on up to 74 Heron qubits and cross-validated against Quantinuum trapped-ion hardware, finding that classical methods diverged from the quantum result around cycle 15 at 51 qubits.
- Algorithmiq (27 July, announced with IBM 30 July) estimated operator Loschmidt echoes on 56 Heron qubits, simulating heterogeneous quantum matter relevant to catalysts and battery electrolytes. Its contribution was a framework for *trusting* a quantum result with no classical ground truth to compare against: build a model of the device's noise, inject controlled noise across multiple processors to show the result is stable, and report unbiased error mitigation with quantified uncertainty. Algorithmiq also open-sourced its classical simulator, "monoprop," so that others could attack the claim. IBM notes the result stood unchallenged on its Quantum Advantage Tracker for eight months.

Independent analysis was appropriately unimpressed by the packaging and impressed by the substance. The three papers make *different kinds* of claim — a complexity-theoretic argument, an empirical outperformance, and a trust protocol — and IBM marketed all three as one "era." What is genuinely new is the shift from unverifiable assertion to statistically testable certificates under stated assumptions. What is not new is any commercial capability, and none of it moves cryptographic timelines.

2.2 The logical qubit race

SYSTEM	LOGICAL QUBITS	PHYSICAL QUBITS	RATIO	DATE
QuEra	96	448	4.7:1	Jan 2026
Quantinuum Helios	48	98	2.0:1	Nov 2025
Atom Computing + Microsoft	24	1,180	49.2:1	Nov 2024
Infleqtion	12	1,600	133:1	Dec 2025
Google Willow	1	105	105:1	Dec 2024

Figures per a leaderboard requiring peer review, verifiable preprint with public code, or pre-registered announcement, and requiring that logical qubits be operated as logical qubits rather than passively encoded. Vendor roadmap targets are excluded. Ratios reflect different code distances and protection levels and are not directly comparable.

Verified logical qubits, and the physical qubits each one costs

Bar length is logical qubit count. The ratio at right is the number that decides when fault tolerance arrives.

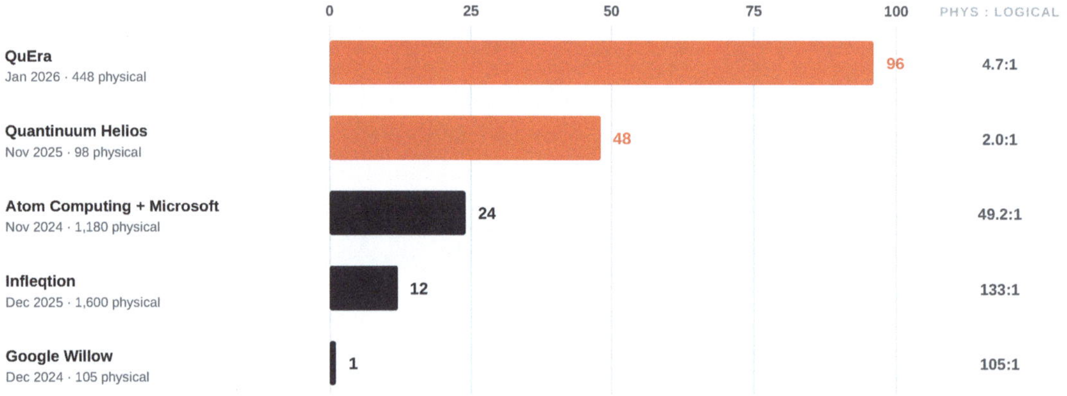

Figure 2. Ratios encode different code distances and protection levels and are not directly comparable; the leaderboard requires logical qubits to be operated, not passively encoded, and excludes roadmap targets. *Source: Quantum Zeitgeist verified logical qubit leaderboard, 2026; company technical disclosures.*

Two things stand out. First, neutral atoms and trapped ions are currently ahead on this metric while superconducting leads on speed and depth — reinforcing the platform-convergence point. Second, Quantinuum's near "five-nines" logical fidelity on Helios, achieved with a new QEC code family, is arguably a more meaningful result than any raw count: it is the first credible sighting of logical error rates in the range fault tolerance actually needs.

Supporting results were as important as the headline numbers. QuEra, Harvard and MIT demonstrated logical-level magic state distillation on neutral atoms — the missing piece for a universal gate set, since Clifford operations alone are classically simulable. Peer-reviewed QEC papers rose from 36 in 2024 to roughly 120 between January and October 2025, and all seven principal code families were implemented on hardware. Oxford Ionics reported 99.99% two-qubit gate fidelity before IonQ acquired it.

2.3 Hardware and architecture

IBM unveiled Nighthawk in November 2025: 120 qubits in a square lattice with roughly 16× greater effective circuit depth than its predecessor, and announced a shift to 300mm wafer fabrication. Its roadmap runs Loon (2025, c-couplers enabling high-rate qLDPC codes), Kookaburra (2026, first module combining qLDPC memory with a logical processing unit), Cockatoo (2027, inter-module entanglement via universal adapters), Starling (2028–29, magic state injection across modules, then 100 million gates on 200 logical qubits), with Nighthawk versions reaching 15,000-gate circuits and nine connected modules totalling 1,080 qubits by 2028. The error-correction approach is bivariate bicycle codes — the [[144,12,12]] "gross" code encoding 12 logical qubits in 144 data qubits, and a [[288,12,18]] "two-gross" variant.

Quantinuum launched Helios in November 2025 — 98 physical qubits, 48 logical, 99.921% average two-qubit fidelity — then IPO'd in June 2026, raising $1.7 billion gross and ending Q2 with $2.1 billion in cash. Q2 2026 revenue was $8 million, up 279% year over year, against a $597 million GAAP net loss and negative GAAP gross margin. Full-year guidance is $28–32 million. Sol is on track for 2027 with its trap chip back from fabrication; Apollo, a fully fault-tolerant universal system, targets 2029.

Neutral atom academic records. Caltech's 6,100-atom array (September 2025, *Nature*) held 13-second coherence with 99.98% single-qubit fidelity. Harvard and MIT demonstrated continuous operation of a coherent 3,000-qubit system, removing the reload-and-restart ceiling.

China. Zuchongzhi 3.2 (published December 2025, *Physical Review Letters*) is a 107-qubit superconducting processor that achieved below-threshold error correction at surface-code distance seven with an error-suppression factor of 1.4, using an all-microwave leakage-suppression architecture distinct from Google's approach. Origin Quantum launched Origin Wukong-180 in May 2026 and China opened a superconducting machine for commercial access in October 2025.

Microsoft's Majorana 2 (June 2026) reported parity lifetimes around 22 seconds, a topological gap roughly doubled to ~70 μeV, and materials changes from aluminium to lead with indium arsenide semiconductor redesign — amid continuing community scepticism about the underlying physics claims.

NVQLink (October 2025) established the GPU-QPU interconnect standard, with 17 hardware builders and nine national labs, and NVIDIA shipped `cudaq-realtime` for programming logical QPUs in March 2026.

2.4 Money, consolidation, and policy

Public commitments to quantum reached $56.7 billion cumulatively, up $12.7 billion in a year. Venture capital hit $4.9 billion in 2025, more than double the prior year, with over $2.7 billion into US-headquartered companies. The QED-C put the total quantum market at $1.9 billion for 2025 — $1.4 billion computing, $470 million sensing — projecting above $4 billion by 2028, with computing growing about 30% annually to $3 billion and sensing about 32% to $1.1 billion. There were 7,418 quantum-engaged organizations globally, 556 of them pure-play, employing roughly 16,500 people.

National programmes: the US National Quantum Initiative committed $1.2 billion for 2019–2024 and $1.8 billion for 2025–2029 across five DOE research centres. China's spend is estimated between $5 billion and $15 billion with genuine dispute over disbursement. The UK committed £2.5 billion over ten years plus a further £2 billion announced March 2026. Germany has a €2 billion national programme; France €1.8 billion; Japan roughly $900 million within a broader R&D budget, having designated 2025 the "first year of quantum industrialization"; South Korea $2.3 billion by 2035; Australia over AU$2.3 billion; India roughly $1 billion; and the EU Quantum Flagship €1 billion across 2018–2028.

Consolidation accelerated. IonQ acquired Oxford Ionics for approximately $1.075 billion and Lightsynq; Google acquired Atlantic Quantum. Valuations reached $10 billion for Quantinuum pre-IPO, $7 billion for PsiQuantum, $5.75 billion for SandboxAQ. QuEra raised $232 million in February 2025 from investors including Google Quantum AI and SoftBank Vision Fund 2.

The US signed two executive orders on 22 June 2026. "Ushering in the Next Frontier of Quantum Innovation" created the Quantum Computer for Application Development and Discovery Science initiative, targeting a scientifically capable quantum computer by 2028. "Securing the Nation Against Advanced Cryptographic Attacks" set hard deadlines: PQC migration leads designated within 30 days, OMB guidance at 90 days, a NIST pilot and updated National Quantum Strategy at 180 days, CISA cryptographic bill-of-materials guidance at 270 days, PQC for key establishment on high-value assets by 31 December 2030, and PQC digital signatures by 31 December 2031. DARPA's Quantum Benchmarking Initiative continues to target a useful $1 billion-class quantum computer by 2033.

2.5 The counter-narrative

Three developments cut against the momentum, and an honest reading has to hold them at the same time.

Valuations detached from revenue. Insiders at IonQ, Rigetti and D-Wave sold over $988 million in stock since mid-2021 against roughly $4.2 million in purchases. Pure-play quantum companies traded at price-to-sales ratios between 60 and 459; no company at the front of a genuinely transformative technology has historically sustained above 30. Quantinuum, the revenue leader among pure-plays, booked $8 million in a quarter. The technology is real; the current prices embed timelines that the technology's own roadmaps do not support.

A replication problem. In August 2026, *New Scientist* reported that quantum computing research may be facing a replication crisis — results that do not survive independent reproduction. This is not a scandal so much as a predictable consequence of a field where hardware is scarce, calibration is idiosyncratic, and classical baselines are frequently under-optimized. It is also precisely why the verification protocols from Algorithmiq, Qedma and IBM matter more than any qubit count.

A talent floor. Roughly 600 to 700 quantum-error-correction specialists exist worldwide against an estimated 2030 need of 5,000 to 16,000, with training pipelines up to a decade long. The global pure-play workforce of ~16,500 grew by only about 2,000 in a year. Hardware roadmaps are, in part, hiring plans that have not yet been satisfied.

Chapter 3 — Who Is at the Forefront

Leadership in quantum computing is not a single ranking. A company can lead on fidelity and trail on scale, or lead on error correction and have no revenue. The honest framing is: who leads at *what.*

3.1 The full-stack incumbents

IBM — superconducting. The most detailed and most publicly tracked roadmap in the industry, and the only one that has repeatedly hit its own dates. Nighthawk at 120 qubits, Loon proving qLDPC connectivity, Kookaburra in 2026, Starling in 2029 at 200 logical qubits and 100 million gates, built in Poughkeepsie. Its 2026 advantage claims are the most rigorously verified in the field. CEO Arvind Krishna predicted first real-world quantum advantage in 2026 and IBM published three papers arguing it delivered. Strengths: manufacturing depth, 300mm wafer transition, software stack, error-correction theory. Weakness: superconducting connectivity constraints, and a company-wide incentive to declare milestones.

Google Quantum AI — superconducting, now also neutral atoms. Achieved the field's foundational below-threshold error-correction result with Willow in December 2024 and the first verifiable advantage with Quantum Echoes in October 2025. Its algorithmic work has arguably reshaped the field more than its hardware: Gidney's RSA revision and the March 2026 elliptic-curve paper both came from Google. Opened a Boulder neutral-atom lab in March 2026, acquired Atlantic Quantum, and is migrating its own infrastructure to post-quantum cryptography by 2029 — a signal worth more than any press release. Targets commercially useful systems by around 2030.

Microsoft — topological, plus Azure Quantum as a platform. The highest-risk, highest-payoff hardware bet in the industry. Majorana 2's 22-second parity lifetimes and doubled topological gap are real engineering progress; whether they are progress toward a topological qubit depends on physics that the community has not settled. Microsoft's platform position — orchestrating other people's hardware, plus the Atom Computing partnership that produced 24 logical qubits — is commercially sound regardless of how the topological bet resolves.

NVIDIA — the classical half. Not a qubit company, and arguably the most strategically positioned player in the sector. NVQLink makes GPUs the mandatory control and decoding layer for fault-tolerant quantum computing across 17 hardware builders. CUDA-Q and `cudaq-realtime` are becoming the default programming surface. Whoever wins on qubits, NVIDIA sells the rack it sits in.

3.2 The pure-play leaders

Quantinuum — trapped ions. The clearest technical leader on quality, and now the sector's most significant public company after its June 2026 IPO raised $1.7 billion. Helios at 98 physical / 48 logical qubits with 99.921% two-qubit fidelity and near five-nines logical fidelity. Sol in 2027, Apollo in 2029, universal fault tolerance targeted by 2030. Commercial traction is genuine but small: $8 million quarterly revenue, 180 organizations on Nexus, Oracle Cloud and HPE partnerships, JPMorgan Chase as a named optimization customer. Weakness: trapped-ion gate speed, and a scaling story that depends on photonic interconnect between traps.

IonQ — trapped ions, aggressively acquisitive. Bought Oxford Ionics for approximately $1.075 billion and Lightsynq for photonic interconnect. Its published roadmap is the most aggressive in the industry: 10,000 physical qubits on a single chip in 2027, 20,000 across two linked chips in 2028, over 2,000,000 physical qubits by 2030 yielding 40,000–80,000 logical qubits with shallow-memory logical error rates below 1e-12. Oxford Ionics' 99.99% two-qubit fidelity is a genuine asset. Treat the 2030 figures as targets, not forecasts — they imply a two-hundred-fold jump in three years.

QuEra — neutral atoms. Current logical-qubit leader at 96 from 448 physical, with logical magic state distillation demonstrated alongside Harvard and MIT. Aquila (256-qubit analog) runs on Amazon Braket; Gemini gate-model units shipped to Japan's AIST and the UK NQCC. Roadmap targets 100 logical qubits on ~10,000 physical. Backed by $232 million including Google Quantum AI and SoftBank.

PsiQuantum — photonics, $7 billion valuation. The most all-or-nothing architecture: skip the intermediate-scale era entirely and build a million-qubit fault-tolerant machine using semiconductor foundry processes, with utility-scale sites announced in Brisbane and Chicago. Its million-qubit timeline has slipped from 2025 to 2027, which is the relevant datapoint.

Pasqal, Atom Computing, Infleqtion — neutral atoms. Pasqal demonstrated solving differential equations with logical neutral-atom qubits. Atom Computing's Microsoft partnership produced the 24-logical-qubit result. Infleqtion reported 12 logical qubits from 1,600 physical and is also a leading quantum-sensing commercial player.

Alice & Bob, Rigetti, IQM, Oxford Quantum Circuits, Quantum Circuits Inc. — superconducting specialists. Alice & Bob's cat-qubit approach targets bias-noise protection that could cut error-correction overhead substantially. Rigetti is a public company with modest revenue and a full-stack superconducting offering. IQM leads European on-premises deployment.

D-Wave — quantum annealing. Commercially the earliest revenue in the sector and technically the most contested: annealing is not universal quantum computation and its advantage claims have been repeatedly matched classically. Real customers, real deployments, narrow applicability.

Xanadu, Quandela, ORCA — photonics. Xanadu's Christian Weedbrook predicted 2026 would bring integration of early fault-tolerant photonic building blocks; the company signed a quantum-education MOU with the University of Guelph in August 2026.

Alice & Bob, Diraq, Silicon Quantum Computing, Quantum Motion — silicon spin qubits and cat qubits. The silicon-spin thesis is that CMOS manufacturing eventually wins on cost per qubit; the platform trails on demonstrated logical qubits.

3.3 China

Chinese capability is consistently misjudged in both directions. Zuchongzhi 3.2 is a genuine below-threshold error-correction result on 107 superconducting qubits, published in a top-tier Western journal, using an architecture distinct from Google's — that is not a copy. USTC's photonic and superconducting programmes are world class. Origin

Quantum ships commercial systems, most recently Origin Wukong-180 in May 2026, and China opened a superconducting machine for commercial access in October 2025. Baidu has contributed platform work.

At the same time, published Chinese logical-qubit results trail Quantinuum and QuEra, the funding figures are unverifiable, and export controls on cryogenics, lasers and control electronics are a real constraint. The reasonable assessment: China is a peer competitor on superconducting hardware and photonics, behind on error-corrected systems integration, and opaque enough that confident rankings are unwise.

3.4 The enablers

The supply chain is where quiet fortunes are being made: Quantum Machines and Zurich Instruments in control electronics, Bluefors and Oxford Instruments in cryogenics, Riverlane in error-correction decoding, Q-CTRL in error suppression, Qedma and Algorithmiq in error mitigation and verification, SandboxAQ at $5.75 billion in quantum-adjacent security and simulation software, and QpiAI, which opened a quantum chip foundry in India in August 2026 targeting 10,000-qubit processors. Aquark's quantum clock trialled with the Royal Navy and Saab's quantum-timed radar network demonstration in August 2026 are reminders that quantum *sensing* is commercializing years ahead of quantum computing.

Chapter 4 — What to Expect by August 2027

The honest way to project one year out is to separate what is already fabricated and scheduled from what is aspirational, and to state which claims would falsify the projection.

Published roadmaps, 2025–2030

Company targets, not independent forecasts. The 2027 column is the near-term test of every line below.

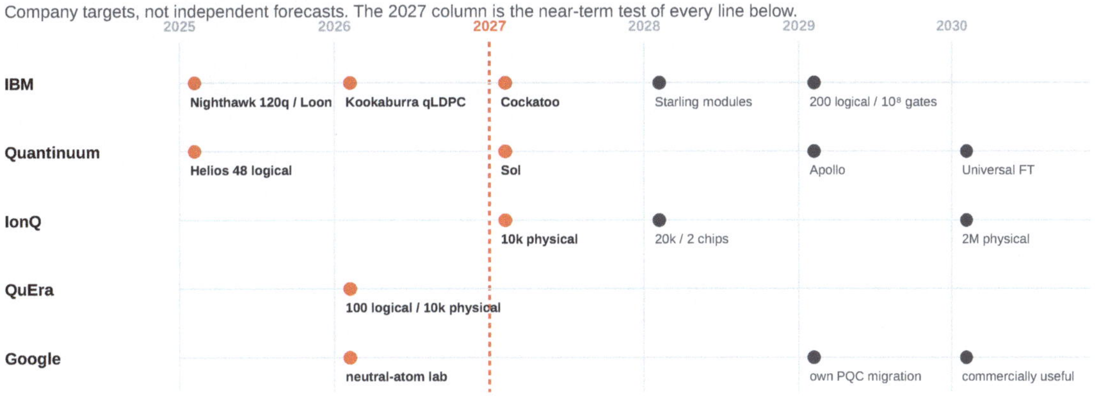

Figure 3. Milestones through 2027 are shown in accent because they are checkable inside the projection horizon of this volume. Every figure on this chart originates with the company that published it. *Source: IBM Quantum roadmap; Quantinuum, IonQ and QuEra published roadmaps; Google Quantum AI statements.*

4.1 Hardware: high confidence

Expect IBM's Kookaburra — the first module to store information in qLDPC memory and process it with a logical processing unit — to be demonstrated, and Cockatoo work on inter-module entanglement via universal adapters to be underway. Expect Nighthawk iterations approaching 7,500-gate circuits on the path to 15,000 by 2028. IBM has hit its published dates consistently enough that the base rate favours delivery.

Expect Quantinuum's Sol to launch; the trap chip is back from fabrication and in product validation. Expect logical qubit counts on trapped-ion and neutral-atom platforms in the range of 100 to a few hundred, with the physical-to-logical ratio continuing to compress. QuEra's 100-logical-qubit target on roughly 10,000 physical qubits is plausible on its stated schedule.

Expect neutral-atom arrays above 10,000 physical atoms in academic settings, given Caltech's 6,100 and the pace of optical-tweezer scaling. Expect IonQ to report progress toward 10,000 physical qubits on a chip; treat the full figure as a target.

Practical translation of "power" in August 2027: on the leading systems, circuits of order 10^4 to 10^5 error-corrected operations on tens to low hundreds of logical qubits, with the best logical error rates approaching 10^-5 to

10^-6. That is enough for genuinely novel physics simulation on carefully chosen Hamiltonians. It is not enough for commercial chemistry, not enough for optimization advantage, and roughly five orders of magnitude short of breaking RSA-2048.

4.2 Advantage claims: moderate confidence

Expect a small number — three to ten — of new quantum-advantage claims, with the verification framing now standard rather than optional. Expect at least one high-profile claim to be classically dequantized within the year; that is healthy, and the Algorithmiq open-source simulator was released precisely to make it happen faster.

Expect the centre of gravity to be quantum simulation of condensed matter, spin dynamics, and lattice gauge theory rather than optimization or machine learning. This is where the mathematics genuinely favours quantum hardware.

Expect the replication conversation to intensify, and expect it to produce something useful: pre-registration norms, mandatory classical baselines, shared benchmark suites. Riverlane's QuOps framing or something like it will likely gain traction as the industry's shared yardstick.

4.3 Cryptography and policy: high confidence

Expect PQC migration to shift decisively from voluntary to mandatory. The June 2026 executive order deadlines land through 2027: OMB guidance, the NIST pilot, an updated National Quantum Strategy, and CISA cryptographic bill-of-materials guidance all fall inside the window. Expect cryptographic inventory — knowing what algorithms are used where — to become an audit item rather than an engineering nicety.

Expect further downward revisions to the resource estimates for breaking RSA and ECC. Three papers in ten months cut requirements by two orders of magnitude through algorithmic work alone, and there is no reason to think that vein is exhausted. Iceberg Quantum's Pinnacle architecture claim of under 100,000 physical qubits for RSA-2048, using quantum LDPC codes, remains simulation-validated only — watch for hardware corroboration or refutation.

Expect no cryptographically relevant quantum computer by August 2027. Nothing in any credible roadmap puts one there.

4.4 Commercial: high confidence in the negative

Expect no commercial quantum advantage in drug discovery, portfolio optimization, logistics, or machine learning on classical data. The pharmaceutical timeline in the literature is 20–100 logical qubits for early fault tolerance in 2027–2030, 200–500 for small-molecule simulation in 2030–2033, and 2,000-plus for true advantage in 2032–2036. Quantum machine learning has no result on ordinary business data that commands broad acceptance, and dequantization work has systematically closed the gaps that looked exponential.

Expect the market to reach roughly $2.5 to $3 billion in total quantum revenue, on track for the QED-C's $4 billion-plus by 2028, with sensing continuing to outgrow expectations. Expect more consolidation — the sector has too many sub-scale hardware companies for its revenue base. Expect at least one significant correction in quantum equities; price-to-sales ratios between 60 and 459 do not survive contact with $8 million quarters indefinitely.

Expect hybrid quantum-HPC-GPU deployment to become the default architecture, with more Quantinuum-Oracle-style placements of QPUs inside AI data centres.

4.5 What would falsify this

If any of the following happens by August 2027, the projection above is wrong and the timeline should be compressed: a demonstration of Shor's algorithm factoring a previously unknown RSA key of non-trivial size; hardware validation of the sub-100,000-qubit RSA architecture; a logical qubit count above 1,000 with operational gates; or a quantum result on classical business data that survives six months of classical counterattack. Fred Chong of the University of Chicago specifically flagged the possibility of realistic Shor implementations on unknown keys as a 2026 development to watch.

Conversely, if Kookaburra slips, if Sol slips, or if a major advantage claim collapses under dequantization, the 2029–2030 fault-tolerance consensus should be read as 2033-plus.

Chapter 5 — Industry Impact

5.1 The honest structure of the opportunity

Quantum computing will not arrive as a general productivity improvement. It will arrive as a narrow, extremely valuable capability in a small number of problem classes, and the discipline required is to know which class you are in.

Three categories, in descending order of confidence:

Simulating quantum systems. Chemistry, materials, catalysis, and nuclear and particle physics. The problem *is* quantum, so the mapping to quantum hardware is natural rather than contrived. This is where advantage is most likely and where the early scientific results already are.

Structured mathematical problems with known quantum algorithms. Factoring, discrete logarithms, certain linear algebra with strong preconditions. Enormous consequences, very narrow applicability, and in the cryptographic case the consequence is destructive rather than productive.

Optimization and machine learning. The largest addressable market and the weakest evidence. Most claimed speedups are quadratic at best, which error-correction overhead can erase entirely, and dequantization has repeatedly closed apparent exponential gaps.

5.2 Pharmaceuticals and life sciences

Activity is real and outcomes are not yet. Boehringer Ingelheim works with Google Quantum AI on molecular dynamics and protein-ligand modelling. AstraZeneca with IonQ, AWS and NVIDIA has demonstrated acceleration in catalytic chemistry workflows. Moderna and IBM are exploring hybrid AI-quantum approaches for mRNA design. QSimulate reports up to 1,000× speedups in quantum-enhanced calculations with top-20 pharmaceutical clients — a vendor figure, and one where the baseline matters enormously. Patent filings from Merck, Roche, Pfizer and Johnson & Johnson indicate substantial internal programmes.

What has actually been achieved is the first chemically meaningful hybrid workflows — quantum subroutines embedded in classical pipelines, producing results a chemist would recognize as valid, on molecules classical methods could already handle. The published expectation is 20–100 logical qubits in 2027–2030, 200–500 enabling small-molecule simulation in 2030–2033, and 2,000-plus for genuine advantage in 2032–2036.

The strategic argument for acting now is not computational; it is organizational. Building the talent, the workflow integration, and the problem-formulation capability takes years, and the companies that have those assets when the hardware arrives will convert first. The counter-argument is equally valid: a pharmaceutical company that spends nothing on quantum until 2030 and then buys capability may end up ahead of one that spent a decade on proofs of concept. The defensible middle is a small, senior, well-scoped team whose job is readiness rather than results.

5.3 Chemicals, materials and energy

The same physics, a shorter path to value, and less attention. Catalyst design, battery electrolytes, nitrogen fixation, carbon capture sorbents, and high-temperature superconductors are all problems where classical density functional theory makes approximations that quantum simulation could eventually remove. The Algorithmiq–IBM result on heterogeneous quantum matter is explicitly framed around catalysts and battery electrolytes. The US Department of Energy committed $30 million through ARPA-E for quantum computing applied to chemistry and materials simulation.

Energy is where analysts see the earliest structural fit, partly because the sector is simultaneously being reshaped by compute-driven demand and already runs large-scale simulation as a core function.

5.4 Financial services

Banks are the most active and the most candid. JPMorgan Chase is a named Quantinuum customer working on constrained quantum optimization, and its quantum lead has pointed to portfolio optimization, cryptography and risk modelling, citing exponential space advantage in streaming algorithms — a real theoretical result with narrow application. HSBC and IBM's September 2025 bond-trading result showed advantages only on noisy hardware, for reasons not fully explained. More than fifteen global banks have active programmes.

The candid industry framing is that quantum cannot do much for banking yet, and banks cannot afford to wait — because the cryptographic exposure is immediate and the option value on optimization is worth a small standing team. Derivative pricing via quantum amplitude estimation offers a quadratic speedup, which is real but modest once error-correction overhead is counted.

5.5 Logistics, manufacturing and operations

This is where the gap between vendor pitch and evidence is widest, and where an operations audience should be most sceptical.

Routing, scheduling, bin packing, and network design are NP-hard, and quantum algorithms for them offer at best quadratic speedups with heavy preconditions on problem structure and data loading. Classical solvers — commercial mixed-integer programming, metaheuristics, and increasingly learned heuristics — have improved enormously and are the correct baseline. Most published "quantum optimization wins" compare against a weak classical baseline or run on a simulator.

The defensible operational position: quantum optimization is emerging, not deployed, and nothing in the next five years should displace investment in data quality, process discipline, and classical solver capability. The organizations that will benefit from quantum optimization in the 2030s are the ones whose problems are cleanly formulated and whose data is trustworthy — which is exactly the work that pays off classically today. Quantum readiness and operational excellence are the same programme.

5.6 Artificial intelligence

The convergence is real in one direction and largely aspirational in the other.

Classical AI is materially improving quantum computing: machine-learned decoders for error correction, reinforcement learning for pulse calibration, and neural-network methods for classical simulation baselines. NVIDIA's positioning depends on this.

Quantum computing improving AI is a different matter. Quantum machine learning has demonstrated genuine advantage on *quantum* data — a 2025 photonic result achieved roughly 11.8 orders of magnitude fewer samples than classical methods for learning specific quantum processes. On classical business data, no result commands broad acceptance. In one peer-reviewed study, ordinary logistic regression beat quantum classifiers on three of five datasets. A May 2026 preprint claiming quantum efficiency gains ran on simulators and took roughly 1,200 times longer than the classical alternative. Three structural bottlenecks persist: data loading can cost as much as the computation, most speedups are quadratic and vanish under error-correction overhead, and dequantization keeps closing apparent exponential gaps.

Quantum large language models are not a near-term prospect. The realistic near-term intersection is scientific machine learning trained on quantum-simulation output.

5.7 Sensing: the part that is already commercial

Quantum sensing generated $470 million in 2025 and is growing at roughly 32% annually toward $1.1 billion by 2028 — smaller than computing but far closer to deployed reality. Aquark's quantum clock trialled with the Royal Navy and Saab's quantum-timed radar network demonstration, both in August 2026, are operational trials rather than laboratory results. Magnetometry for biomedical imaging, gravimetry for subsurface survey, and atomic clocks for GPS-denied navigation are commercializing now.

For an executive audience this is the most underrated line in the whole sector: the quantum technology most likely to touch your operations before 2030 is a sensor, not a computer.

Chapter 6 — Threats

6.1 The cryptographic threat, stated precisely

Almost all public-key cryptography in use today rests on two problems: factoring large integers (RSA) and computing discrete logarithms on elliptic curves (ECDSA, ECDH, and the key exchange protecting essentially all web traffic). Shor's algorithm solves both efficiently on a sufficiently large fault-tolerant quantum computer. Symmetric cryptography — AES — is far less exposed; Grover's algorithm offers a quadratic speedup, addressed by doubling key length.

The threat is therefore specific: confidentiality and authenticity of anything protected by RSA or elliptic-curve cryptography. That includes TLS, code signing, VPNs, secure boot, most blockchain wallets, and the certificate hierarchy underneath all of it.

6.2 Harvest now, decrypt later

The threat does not begin when a quantum computer exists. It began when adversaries started recording encrypted traffic to decrypt later. Any data with a confidentiality requirement extending beyond the arrival of a cryptographically relevant quantum computer is already exposed if it crosses a network an adversary can observe.

The relevant calculation, sometimes called Mosca's inequality, is simple: if the time your data must stay secret plus the time it takes you to migrate exceeds the time until a capable quantum computer exists, you are already late. For genomic data, classified material, long-term contracts, and intellectual property with multi-decade value, the answer is unambiguous.

6.3 How fast the estimates are moving

This is the most alarming trend in the field, and it has nothing to do with hardware.

- **2019:** Gidney and Ekerå estimate RSA-2048 requires roughly 20 million noisy physical qubits.
- **May 2025:** Gidney revises to fewer than one million physical qubits in under a week of runtime — a twenty-fold reduction achieved through approximate residue arithmetic and improved magic state cultivation. Algorithms, not hardware.
- **February 2026:** Iceberg Quantum's Pinnacle architecture, using quantum LDPC codes, claims fewer than 100,000 physical qubits — another ten-fold cut. It requires connectivity beyond nearest-neighbour and is validated in simulation only.
- **March 2026:** Google Quantum AI, with Justin Drake and Dan Boneh, shows 256-bit elliptic-curve cryptography — the cryptography protecting most cryptocurrency — falls to fewer than 500,000 physical qubits with runtime in minutes rather than days. Roughly a twenty-fold improvement. Google withheld the attack circuits and published a zero-knowledge proof verifying the estimates instead.

Estimated physical qubits needed to break deployed public-key cryptography

Two orders of magnitude removed in ten months — through algorithms, not hardware. Log scale.

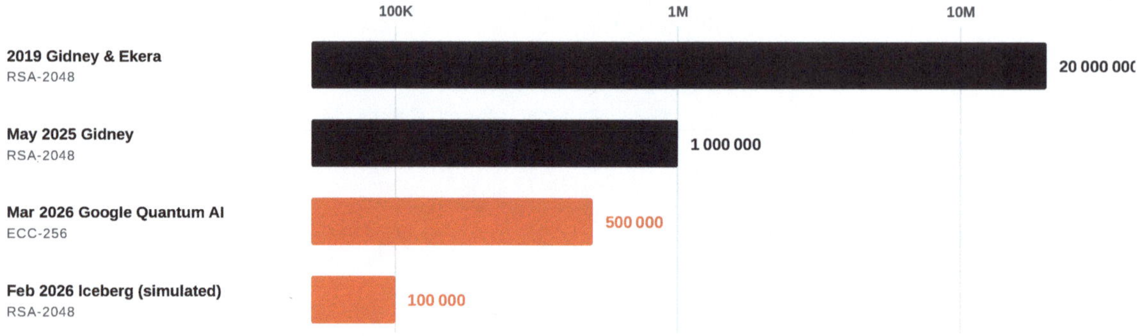

Figure 1. The cryptographic threat timeline compressed twice in ten months without any change in hardware capability. The Iceberg figure is validated in simulation only and requires connectivity beyond nearest-neighbour. *Source: Gidney & Ekerå 2019; Gidney, May 2025; Iceberg Quantum, Feb 2026; Google Quantum AI, Mar 2026, as reported by The Quantum Insider.*

Two orders of magnitude in ten months, entirely from mathematics. Anyone whose migration plan was built on the twenty-million-qubit figure is planning against a number that no longer exists.

6.4 What credible people now say about timing

Scott Aaronson has been the field's most reliable check on hype for twenty years. On 30 April 2026 he wrote that hardware and error-correction experts he trusts believe a fault-tolerant quantum computer able to break deployed cryptosystems ought to be possible around 2029, deferred explicitly to their judgment over his own, and framed the post as a warning so that nobody could later say they were not told.

Market participants closest to the hardware are acting accordingly. Google targets full PQC migration by 2029, as does Cloudflare — four to six years ahead of regulatory requirements. Ethereum researcher Justin Drake, a co-author on the elliptic-curve paper, has said his confidence in quantum-capable machines by 2032 rose significantly. Bitcoin developer Bit Paine put roughly ten years as most likely while assigning an uncomfortably high probability to something disruptive within five.

Set against this, the December 2023 Global Risk Institute survey of 37 experts put the probability of a cryptographic threat within five years at under 1%, with the consensus tipping point at 10 to 20 years. That survey predates all three of the papers above. Surveys lag algorithms.

The regulatory calendar has effectively settled the argument: NSA CNSA 2.0 requires quantum-resistant support in new national security system acquisitions from 2027; the US executive order mandates PQC key establishment on high-value assets by end of 2030 and signatures by end of 2031; EU high-risk infrastructure and UK high-priority systems transition around 2030–2031; Microsoft targets full transition by 2033; NIST IR 8547 disallows RSA-2048 and ECC-256 by 2035.

6.5 The non-cryptographic threats

Concentration of capability. A machine that can simulate catalysis or design materials better than anyone else is a durable economic advantage held by whoever owns it. Access is currently mediated by cloud platforms, which is healthier than exclusive ownership, but the assets are extraordinarily capital-intensive and the roadmaps imply consolidation.

Geopolitical asymmetry. Export controls on cryogenics, lasers and control electronics are already instruments of policy. A capability gap in cryptanalysis between state actors is destabilizing in ways that are hard to model and harder to verify, precisely because a successful cryptanalytic capability is the last thing a state would announce.

Hype-driven misallocation. Price-to-sales ratios between 60 and 459, insiders selling $988 million against $4 million of buying, and $8 million quarterly revenues at the sector's revenue leader describe a market pricing 2035 outcomes into 2026 equities. A correction would be healthy for the market and damaging for the research programmes it funds — the deeper risk of a bust is not lost capital but lost decade.

Scientific credibility. The replication concerns reported in August 2026 are the field's most self-inflicted risk. A discipline whose flagship claims do not reproduce forfeits the public trust that funds it. The counterweight is the verification turn described in Chapter 2, and it needs to hold.

Talent bottleneck as a security issue. Six to seven hundred error-correction specialists worldwide is a small enough population that hiring is a geopolitical event.

Chapter 7 — Benefits

7.1 The case that justifies the investment

Strip away the noise and quantum computing's benefit case rests on one claim: nature is quantum mechanical, and we currently cannot compute what nature does.

Every material, every chemical reaction, every enzyme, every superconductor is a many-body quantum system. Classical simulation of such systems requires approximations — density functional theory, coupled cluster, Monte Carlo with sign-problem workarounds — that work well until they do not, and it is precisely the cases where they fail that are scientifically and commercially interesting. Strongly correlated electron systems. Transition-metal catalysis. High-temperature superconductivity. Nuclear matter at neutron-star densities.

A fault-tolerant quantum computer with a few thousand good logical qubits would let us compute these directly. That is not a productivity improvement; it is a new instrument, in the sense that the telescope and the microscope were instruments. The economic value is unbounded in the way that the value of a new instrument always is, and equally unforecastable.

7.2 Concrete near-term benefits

Chemistry and catalysis. Direct simulation of reaction pathways would compress the design cycle for fertilizer catalysts, carbon capture sorbents, and industrial processes that currently consume single-digit percentages of global energy. Nitrogen fixation alone is roughly 1–2% of world energy use.

Materials. Battery electrolytes and cathode chemistry, high-temperature superconductors, and photovoltaic materials all hinge on correlated-electron physics. The Algorithmiq–IBM heterogeneous-matter work targets exactly this class.

Drug discovery. Not primarily speed — better prediction. More accurate binding affinity and conformational energetics would raise the hit rate at the stage where failure is cheapest, which matters more than any wall-clock speedup given that the expensive part of pharmaceutical development is clinical, not computational.

Nuclear and particle physics. Real-time dynamics, finite-density matter, and parton structure are inaccessible to classical lattice methods because of the fermion sign problem. This is covered in Chapter 8.

Cryptography, constructively. Quantum key distribution offers information-theoretic security for key exchange over dedicated links, and quantum random number generation provides genuine entropy. Both are deployed today, both are niche, and neither is a substitute for post-quantum cryptography in general-purpose networks.

Sensing, already delivering. Magnetometers sensitive enough for magnetoencephalography without cryogenics, gravimeters for subsurface mapping, optical clocks for GPS-denied navigation and for tests of fundamental physics, and entangled sensor networks for dark matter searches. Commercially available now.

7.3 The benefit nobody markets: better classical algorithms

One of the most valuable outputs of quantum computing research has been classical. Ewin Tang's 2018 dequantization of quantum recommendation systems, and the programme that followed, produced fast classical algorithms that did not previously exist. Tensor-network methods developed to benchmark quantum devices have become standard tools in condensed matter physics. Every serious quantum advantage claim forces the construction of the best possible classical baseline, and those baselines are permanent additions to the toolkit.

This is a real return on a real investment, and it accrues whether or not fault tolerance arrives on schedule.

Chapter 8 — Quantum Computing and the Deep Questions in Physics

This chapter addresses the questions that motivate most public interest in quantum computing, and it does so by drawing a line that popular coverage almost never draws.

A quantum computer is a controllable quantum system. That makes it a laboratory instrument for testing theories of quantum systems. It does not make it a portal, a time machine, or an oracle. When a quantum computer "simulates a wormhole," what has happened is that a mathematical model believed to be dual to a gravitational description has been evolved on qubits. No spacetime was bent. The distinction is not pedantic — it is the whole content of the result.

With that established, the genuine scientific prospects are remarkable.

8.1 Understanding entanglement and superposition

These are not mysteries awaiting solution so much as the operating principles the machines run on. What quantum computers provide is the ability to prepare, manipulate and measure entangled states of a size that was previously impossible, which turns foundational questions into experimental ones.

Entanglement entropy and its scaling, entanglement spectra as diagnostics of thermalization, out-of-time-order correlators as measures of information scrambling — these are now measurable quantities on hardware. Google's Quantum Echoes result is precisely an OTOC measurement, and OTOCs are the central observable in the study of how quantum information spreads and how classical behaviour emerges from quantum substrate. Trapped-ion experiments have used entanglement spectroscopy to identify thermalization signatures in gauge theories.

Where this leads: sharper understanding of decoherence and the quantum-to-classical transition, better tests of whether quantum mechanics is exactly linear at large scale, and quantitative measures of information scrambling in strongly interacting systems. Whether these resolve the *interpretive* question — what superposition means — is doubtful. Interpretation is not obviously an empirical matter.

8.2 Quantum gravity, holography, and the wormhole episode

The AdS/CFT correspondence proposes that certain quantum field theories without gravity are mathematically equivalent to gravitational theories in one more dimension. If true, a quantum computer running the field theory is computing something about a gravitational system. This is the deepest idea in theoretical physics with a plausible route to quantum simulation.

In November 2022, a collaboration including Caltech and Google published in *Nature* a quantum-processor experiment interpreted as observing traversable-wormhole dynamics in the SYK model, with holographic teleportation as the signature. The coverage was extravagant. The scientific response was immediate and substantive:

within months, other physicists argued the specific model used was too simplified to support the holographic interpretation, and that the observed signal admitted a non-gravitational explanation. Peter Woit and others documented the gap between the paper and the press.

The honest summary: a genuine and interesting quantum simulation was performed; the claim that it constituted a wormhole was not established; the episode is a case study in why verification culture matters. Nothing about it moved a gram of mass or bent a metre of spacetime.

What quantum simulation of holography could realistically achieve is progress on the black hole information problem — how information escapes an evaporating black hole — by studying scrambling and recovery in models where both sides of the duality are computable. That is a real research programme with real prospects, and it is a programme in *mathematics and theory*, validated by simulation, not an engineering programme.

8.3 Unification and the fundamental forces

Quantum computers will not derive a unified field theory. Theories come from theorists. What quantum computers can do is *solve* theories we already have but cannot evaluate.

The Standard Model's strong nuclear force is described by quantum chromodynamics, and QCD is only calculable classically in equilibrium, via Euclidean lattice Monte Carlo. Real-time dynamics, finite fermion density, and anything requiring the physical time direction hit the fermion sign problem, where classical importance sampling breaks down. Quantum simulation does not have this problem in principle, because it evolves the system in real time natively.

Hardware demonstrations to date are genuine but small. On IBM superconducting processors: real-time quench dynamics in U(1) and SU(2)/SU(3) lattice gauge theories, parton distribution functions in the Schwinger model, energy-energy correlators in SU(2). On trapped ions at Duke, Maryland, IonQ and Quantinuum: thermalization signatures via entanglement spectroscopy, string breaking, two-hadron scattering, beta-decay transitions. On QuEra's Rydberg arrays: string breaking in quantum link models. A 2026 *Nature Physics* result simulated two-dimensional lattice gauge theories on a qudit quantum computer, and a *Communications Physics* paper laid out a universal framework for simulating Yang-Mills theory.

The resource gap is sobering and should be stated plainly. Full QCD at realistic parameters requires, on optimized block-diagonalization estimates, on the order of 10^27 T gates. Earlier naive encodings implied 10^49 to 10^50 T gates and around 10^11 qubits. Even with the enormous improvements achieved, this is far beyond anything on any roadmap. The field's realistic near-term programme is controlled benchmark problems in lower dimensions and simplified gauge groups, building phenomenological insight and validating methods, with hybrid architectures where classical HPC handles state preparation and measurement analysis while the quantum processor handles time evolution.

Physics questions genuinely in reach on a multi-decade horizon: real-time hadronization and thermalization in heavy-ion collisions; neutron-star matter at densities where the sign problem defeats classical methods; parton distributions and structure functions relevant to collider phenomenology; neutrino-nucleus response functions needed to interpret detector data; and early-universe phase transitions.

8.4 Dark matter and dark energy

Here the leading quantum technology is not the computer. It is the sensor.

Dark matter constitutes roughly 27% of the universe's energy content and has never been directly detected. Leading candidates in the ultralight regime — axions and dark photons — would deposit vanishingly faint microwave signals in resonant cavities at an unknown frequency, forcing a slow scan across a wide band. Standard quantum limits on amplification set the sensitivity floor, and quantum techniques are the only way through it.

This is not speculative. Entangled sensor networks for dark-matter searches were established in *PRX Quantum*. Squeezed states have been shown to enhance photon-axion conversion probability in *Physical Review D*. In-cavity quantum control mitigating the "Rayleigh curse" appeared in *npj Quantum Information*. ORNL published in *Physical Review Research* in February 2026 on two-mode squeezed light in a nonlinear interferometer across distributed optomechanical sensors, using both squeezing and entanglement to reach sensitivities the researchers state are unreachable classically, targeting ultralight dark matter candidates. Fermilab's Yao Lu received a DOE Early Career Award to build a scalable superconducting cavity array linked by remote entanglement, so that many cavities scan as one coordinated system — beginning with a four-cavity prototype. Japanese groups are using quantum computing hardware in dark-matter searches. The DOE's National Quantum Information Science Research Centers run an entangled-haloscope programme.

The mechanism of benefit is concrete: entanglement and squeezing let you evade the standard quantum limit and scan the frequency space faster, which converts directly into experimental reach per year of beam time.

Dark energy is a harder case. It is an accelerating expansion whose cause is unknown, and the leading question — whether it is a cosmological constant or something dynamical — is being attacked observationally by DESI, Euclid and the Vera Rubin Observatory, not computationally. Quantum computation enters the picture through simulation of inflationary and dark-energy cosmological models: there is a genuine literature on quantum computing for inflationary, dark-energy and dark-matter cosmology, and digital quantum simulation of cosmological particle creation has been performed on IBM hardware and published in *Scientific Reports*. A July 2026 theoretical proposal links entropy, dark energy and life within a quantum-gravity framework — interesting, early, and not experimentally established.

The realistic statement: quantum computers will help evaluate candidate models of dark matter and dark energy, and quantum sensors are the primary instruments in the hunt for dark matter particles. Neither will tell us what dark energy is without observational input.

8.5 The history of the universe

Quantum simulation has a natural application to the early universe, because the early universe was a hot, dense, strongly-coupled quantum field theory evolving in real time — exactly the regime classical lattice methods cannot reach.

Digital quantum simulation of cosmological particle creation has been demonstrated on IBM processors. Simulating the electroweak and QCD phase transitions, the dynamics of symmetry breaking, and baryogenesis mechanisms are all identified targets in the gauge-theory literature. Whether the matter-antimatter asymmetry can be explained requires computing real-time dynamics of CP-violating processes in the early universe plasma, which is a sign-problem calculation.

This is one of the most promising long-horizon applications, and one of the least sensationalized. It will not produce a movie of the Big Bang. It may produce a quantitative answer to why there is matter at all.

8.6 Time travel, time dilation, wormholes, and extra dimensions

These deserve a direct answer, because they are the questions readers most want addressed and the ones most abused in coverage.

Time travel. There is no route from quantum computing to time travel. Closed timelike curves are permitted by some solutions of general relativity, and there is a legitimate theoretical literature on what computation would look like if they existed — Deutsch's CTC model, and results on the computational power of postselected closed timelike curves. Some quantum experiments have been described as "simulating" CTCs using postselection and entanglement. These are analogies computed inside ordinary forward-time physics. No experiment has sent information backward in time, and quantum mechanics as currently formulated forbids it. Any claim otherwise is a misreading of a simulation.

Time dilation. This is settled physics, confirmed to extraordinary precision, and quantum technology's contribution is measurement rather than explanation. Optical atomic clocks — quantum devices — measure gravitational time dilation across height differences of a millimetre. The frontier question is whether time dilation interacts with quantum superposition in a way that produces observable decoherence, and whether a clock in superposition of two gravitational potentials behaves as theory predicts. These are experiments people are actually designing, and they probe the boundary between quantum mechanics and general relativity. Quantum computers are not the instrument; quantum clocks and interferometers are.

Wormholes. Covered in 8.2. No wormhole has been created. Traversable wormholes require exotic matter with negative energy density that has never been observed in the required configuration. What quantum computers can do is simulate models conjectured to be dual to wormhole geometries, which is a mathematical statement about a duality, not a statement about spacetime. The value is in what it teaches about information flow and holography.

Extra dimensions. String theory and its relatives require additional compactified dimensions, and no experimental evidence for them exists. Quantum computers can help evaluate dimensionally-reduced field theories and holographic models where an extra dimension appears as an emergent feature of the dual description — the "one more dimension" in AdS/CFT. This is suggestive and mathematically deep. It is not detection. If extra dimensions are found, it will be by particle colliders, precision gravity experiments, or cosmological observation.

A closing note on this section. The reason to be careful here is not scientific squeamishness. It is that the actual prospects are extraordinary enough not to need embellishment. A machine that lets us compute real-time QCD, determine what neutron star cores are made of, and possibly explain why matter exists is one of the most significant instruments humanity could build. Describing it as a time machine makes it sound less impressive, not more, because it makes it sound imaginary.

8.7 Will this help us understand the nature of the universe better?

Yes — in a specific and defensible sense.

Quantum computing extends the range of physical theories we can actually evaluate. For four centuries, physics has advanced through a cycle of theory, calculation, and measurement, and the calculation step has been the bottleneck for the strongly-coupled quantum theories that describe most of the matter in the universe. Removing that bottleneck is a genuine expansion of what is knowable, and it applies to nuclear matter, the early universe, and the structure of the strong force.

It also does something subtler. Building these machines forces the operationalization of concepts that were philosophical — entanglement as a resource with a unit, information as something that scrambles at a measurable rate, error correction as a physical mechanism that reveals which quantum states are robust. Some of the deepest ideas of the past two decades, including the holographic principle and the entropic accounting of black holes, came out of thinking about quantum information rather than about gravity directly.

What it will not do is answer questions that are not computational. Why the constants have the values they do, whether the universe had a cause, what happens inside a singularity, and what a wavefunction *is* are not waiting on more compute.

Chapter 9 — What to Do About It

The purpose of this chapter is to convert everything above into decisions, on the assumption that the reader runs something.

9.1 The cryptographic work is not optional and not future

Inventory first. You cannot migrate what you cannot find. A cryptographic bill of materials — which algorithms, key lengths and certificates are used by which systems, including embedded devices, firmware and third-party dependencies — is the prerequisite for everything else, and CISA guidance on exactly this is due within 270 days of the June 2026 executive order.

Classify by shelf life. Apply the Mosca calculation. Data that must remain confidential past 2035, systems whose firmware cannot be updated in the field, and long-lived signing keys are the priority. Data with a two-year confidentiality requirement is a lower priority than the volume of vendor marketing would suggest.

Sequence: key establishment before signatures. The executive order's own deadlines reflect the risk structure — key establishment on high-value assets by end of 2030, signatures by end of 2031 — because harvest-now-decrypt-later attacks the confidentiality of recorded traffic, while signature forgery requires a live quantum computer.

Make it contractual. Require crypto-agility from vendors now: the ability to swap algorithms without re-architecting. The single most expensive migration failure mode is hard-coded cryptography in systems with fifteen-year service lives.

Assign an owner. The executive order requires federal agencies to designate PQC migration leads within 30 days. Private organizations should do the same, for the same reason: this is a programme, not a project, and it will span more budget cycles than most initiatives survive.

9.2 On the computing side, buy optionality, not capability

Do not buy a quantum computer. Do not fund a large quantum programme. Do fund a small one, and be specific about what it is for.

Identify whether you are in a quantum-relevant industry. If your value creation involves simulating molecules or materials, you are. If it involves optimization, scheduling or forecasting, you probably are not — at least not this decade — and your quantum budget should be spent on classical solver capability and data quality instead.

If you are: formulate problems, don't chase hardware. The scarce skill is not running a quantum circuit; it is translating a business or scientific problem into a form where a quantum algorithm could apply, with an honest classical baseline. That skill transfers to classical work immediately and is the thing you cannot buy in 2032.

Access through cloud, always. Amazon Braket, Azure Quantum, IBM Quantum, Quantinuum Nexus, and increasingly QPUs placed inside AI data centres. There is no defensible reason for a non-research organization to own hardware.

Demand the classical baseline. For every vendor claim, ask five questions: is there reproducible evidence; real hardware or simulator; how strong was the classical baseline; do the advantages survive data loading and error-correction overhead; and is the dataset natural or constructed. Most claims fail on the third question.

9.3 Watch the right indicators

Ignore qubit counts. Track these instead: the physical-to-logical qubit ratio at fixed code distance; whether IBM's Kookaburra and Cockatoo hit their dates; whether Quantinuum's Sol ships in 2027; whether the sub-100,000-qubit RSA architecture gets hardware validation; whether any advantage claim survives twelve months of classical counterattack; and whether logical error rates reach 10^-6.

And watch what the people closest to the hardware do with their own systems. Google migrating its own infrastructure to post-quantum cryptography by 2029, four to six years ahead of any requirement, is the most informative single datapoint in this book.

9.4 The framing that holds up

Quantum computing is an emerging capability, not a deployed one. The correct posture is neither dismissal nor urgency-by-default. It is: treat the cryptographic exposure as a present-tense operational risk with a named owner and a dated plan, and treat the computational opportunity as an option worth a small, senior, honest team.

The organizations that will convert quantum computing into advantage in the 2030s will be the ones whose problems were already well formulated, whose data was already trustworthy, and whose cryptography was already agile. That work pays for itself classically. It is also, not coincidentally, the entire preparation.

Appendix A — Timeline of Key Events, August

2025 to August 2026

DATE	EVENT
Sep 2025	Caltech traps 6,100-atom neutral-atom array; 13 s coherence, 99.98% single-qubit fidelity (*Nature*)
Sep 2025	Harvard/MIT demonstrate continuous operation of a coherent 3,000-qubit system (*Nature*)
Sep 2025	HSBC and IBM report bond-trading result; advantage only on noisy hardware
Oct 2025	Google announces Quantum Echoes on Willow — first verifiable quantum advantage, ~13,000×
Oct 2025	NVIDIA announces NVQLink with 17 QPU builders and 9 US national labs
Oct 2025	China opens a superconducting quantum computer for commercial access
Nov 2025	IBM unveils Nighthawk (120 qubits) and Loon; announces 300mm wafer fabrication
Nov 2025	Quantinuum launches Helios: 98 physical / 48 logical qubits, 99.921% 2Q fidelity
Dec 2025	Infleqtion reports 12 logical qubits from 1,600 physical
Dec 2025	Zuchongzhi 3.2 achieves below-threshold error correction, distance 7 (*PRL*)
Jan 2026	QuEra reports 96 logical qubits from 448 physical — current verified leader
Feb 2026	Iceberg Quantum's Pinnacle architecture claims <100,000 qubits for RSA-2048 (simulation)
Feb 2026	ORNL publishes squeezed-light entangled sensing for ultralight dark matter (*PRR*)
Mar 2026	Google, with Drake and Boneh, shows 256-bit ECC falls to <500,000 qubits in minutes
Mar 2026	Google opens neutral-atom lab in Boulder under Adam Kaufman
Mar 2026	NVIDIA ships `cudaq-realtime` for logical QPU programming

DATE	EVENT
Apr 2026	QED-C reports 2025 quantum market at $1.9B; $56.7B cumulative public funding
Apr 2026	Scott Aaronson warns experts he trusts see cryptographic capability possible by ~2029
Apr 2026	Fermilab's Yao Lu receives DOE Early Career Award for entangled cavity dark-matter array
May 2026	Origin Quantum launches Origin Wukong-180
Jun 2026	Quantinuum IPO raises $1.7B gross
Jun 2026	Microsoft reports Majorana 2: ~22 s parity lifetime, topological gap ~70 μeV
Jun 2026	Two US executive orders on quantum innovation and post-quantum cryptography (22 June)
Jul 2026	IBM/UChicago, Qedma and Algorithmiq publish three quantum advantage claims in four days
Aug 2026	Quantinuum Q2: $8M revenue, +279% YoY; Sol on track for 2027, Apollo 2029
Aug 2026	*New Scientist* reports possible replication crisis in quantum computing research
Aug 2026	Saab demonstrates quantum-timed radar network; Aquark quantum clock trials with Royal Navy
Aug 2026	QpiAI opens quantum chip foundry in India targeting 10,000-qubit processors

Appendix B — Reading the Claims: A Five-Question Filter

Applied to any quantum result, vendor or academic:

1. **Is there reproducible evidence** — peer review, a preprint with public code, or a pre-registered announcement?
2. **Real hardware or simulator?** A simulator result is a claim about an algorithm, not about a machine.
3. **How strong was the classical baseline?** Was a specialist asked to try to beat it, or was a textbook method used?
4. **Do the advantages survive data loading and error-correction overhead?** Many quadratic speedups do not.
5. **Is the dataset natural or artificially constructed?** Constructed hardness is a theorem, not a product.

A sixth, for logical-qubit claims specifically: were the logical qubits *operated* — state preparation, gates, measurement — or merely passively encoded?

Appendix C — Glossary

AdS/CFT — A conjectured equivalence between a gravitational theory in a space with negative curvature and a quantum field theory on its boundary, with one fewer dimension.

Coherence time — How long a qubit maintains its quantum state before environmental interaction destroys it.

CRQC — Cryptographically relevant quantum computer: one large and reliable enough to break deployed public-key cryptography.

Dequantization — Constructing a classical algorithm that matches a claimed quantum speedup, usually by giving the classical algorithm comparable data access.

Fault tolerance — Operating below the error threshold at which adding more physical qubits per logical qubit improves rather than degrades the logical qubit.

Harvest now, decrypt later — Recording encrypted traffic today to decrypt when a quantum computer becomes available.

Logical qubit — An error-corrected qubit encoded across many physical qubits.

Magic state distillation — A procedure for producing high-fidelity non-Clifford resource states, required for universal quantum computation.

Mosca's inequality — If required secrecy duration plus migration time exceeds time-to-CRQC, migration is already late.

OTOC — Out-of-time-order correlator; a measure of how quantum information scrambles, and the observable behind Google's Quantum Echoes.

PQC — Post-quantum cryptography: classical algorithms believed secure against quantum attack, standardized by NIST.

qLDPC codes — Quantum low-density parity-check codes; error-correcting codes with lower overhead than the surface code but requiring longer-range connectivity. IBM's bivariate bicycle codes are an example.

Sign problem — The breakdown of classical Monte Carlo methods for quantum systems at finite fermion density or in real time, due to oscillating-sign contributions.

Superposition — A quantum state that is a weighted combination of basis states, with amplitudes that interfere.

Surface code — The most studied quantum error-correcting code; high overhead, nearest-neighbour connectivity only.

T gate — A non-Clifford gate; T-gate count is the standard measure of a fault-tolerant algorithm's cost.

Appendix D — Sources

All figures in this volume are attributed. Where a number originates with a vendor rather than an independent party, it is labelled as such in the text. Company roadmap targets are identified as targets, not forecasts.

Hardware and error correction

- IBM Quantum, "IBM lays out clear path to fault-tolerant quantum computing" — ibm.com/quantum/blog/large-scale-ftqc
- IBM Quantum roadmap — ibm.com/roadmaps/quantum.pdf
- Tom's Hardware, "IBM unveils new 'Quantum Nighthawk' 120-qubit processor," Nov 2025
- Quantinuum, "Accelerated Roadmap to Universal, Fully Fault-Tolerant Quantum Computing by 2030" (vendor roadmap)
- HPCwire, "Quantinuum Introduces Helios Quantum System as Roadmap Advances Toward Apollo," 5 Nov 2025
- Caltech, "Caltech Team Sets Record with 6,100-Qubit Array," 24 Sep 2025; published in *Nature*
- *Nature*, "Continuous operation of a coherent 3000-qubit system," 2025 — nature.com/articles/s41586-025-09596-6
- Global Times, "China's superconducting quantum prototype 'Zuchongzhi 3.2' achieves key breakthrough," Dec 2025; published in *Physical Review Letters*
- The Quantum Insider, "Microsoft Reports Advances in Majorana 2 Following Debate Over Last Year's Topological Claims," 2 Jun 2026
- Physics (APS), "Experts Weigh in on Microsoft's Topological Qubit Claim"
- Riverlane, "Quantum Error Correction: Our 2025 trends and 2026 predictions"
- Quantum Zeitgeist, "Quantum Logical Qubit Leaderboard 2026"
- QuEra/Harvard/MIT, "Logical-Level Magic State Distillation on a Neutral-Atom Quantum Computer," 2025
- Pasqal, "Solving Differential Equations with Logical Neutral-Atom Qubits"
- IonQ, "IonQ's Accelerated Roadmap" (vendor roadmap); IonQ investor releases on the Oxford Ionics acquisition
- NVIDIA, "NVIDIA Introduces NVQLink," 28 Oct 2025; NVIDIA Quantum blog, "Introducing cudaq-realtime," 16 Mar 2026
- Tech-Insider, "Google Quantum Computing 2026: Neutral Atom Pivot"

Quantum advantage

- PostQuantum, “Google Announces ‘Verifiable Quantum Advantage’ on Willow Quantum Chip” (Quantum Echoes, 22 Oct 2025)
- Quantum Computing Report, “Google Quantum AI Achieves Verifiable Quantum Advantage on Willow with Quantum Echoes”
- IBM Newsroom, “IBM and Algorithmiq Demonstrate Quantum Advantage,” 30 Jul 2026
- The Quantum Insider, “IBM–Algorithmiq Quantum Computation Verification Framework,” 30 Jul 2026
- PostQuantum, “IBM’s Three Quantum Advantage Claims, Fact-Checked”
- C&EN, “IBM claims their quantum computers outperform classical ones,” Aug 2026
- Phys.org, “Quantum computer completes verified task beyond practical reach of classical simulations,” Jul 2026
- The Quantum Insider, “IBM’s Krishna Predicts First Real-World Quantum Advantage in 2026,” 30 Apr 2026

Cryptography and policy

- The Quantum Insider, “Q-Day Just Got Closer: Three Papers in Three Months Are Rewriting the Quantum Threat Timeline,” 31 Mar 2026
- The Quantum Insider, “Post-Quantum Cryptography Timelines: When Will Organizations Migrate?”, 7 Aug 2026
- PostQuantum, “Aaronson Warns: CRQC by 2029 Is Plausible” (Aaronson, Shtetl-Optimized, 30 Apr 2026)
- K&L Gates, “Preparing for Q-Day: New Executive Orders Address Quantum Innovation and Post-Quantum Cryptography,” 26 Jun 2026
- NIST IR 8547; NSA CNSA 2.0; NSM-10
- Palo Alto Networks Cyberpedia, “What Is Q-Day?”
- R Street Institute, “Post-Quantum Cryptography Migration in the United States”
- Global Risk Institute quantum threat timeline expert survey, Dec 2023

Market, funding and industry

- The Quantum Insider, “Global Quantum Computing Market to Double by 2028, Reaching $3 Billion in Revenue, QED-C State of the Global Quantum Industry 2026 Report Finds,” 14 Apr 2026
- The Quantum Insider, “15+ Leading Quantum Computing Countries in 2026,” 26 Mar 2026
- The Quantum Insider, “Quantinuum Revenue Jumps 279% in First Earnings Report Since IPO,” 12 Aug 2026
- Quantinuum, “Quantinuum Reports Second Quarter 2026 Results” (PR Newswire)
- The Motley Fool, “Quantum Computing Stocks IonQ, Rigetti, and D-Wave Are Sending Shockwaves Through Wall Street With This $988 Million Warning,” 22 Jul 2026
- The Quantum Insider, “Top Chinese Quantum Computing Companies in 2026,” 15 May 2026
- The Qubit Report, “China’s Origin Quantum Launches Origin Wukong-180,” 9 May 2026

- The Quantum Insider, “TQI’s Expert Predictions on Quantum Technology in 2026,” 30 Dec 2025
- Dr. Bob Sutor, “Daily Quantum Update for Monday, August 17, 2026”
- Introduction to Quantum Computing for Business, “The timelines: when can we expect useful quantum computers?”
- DARPA Quantum Benchmarking Initiative

Applications

- Drug Discovery World, “Quantum Leap: Is pharma ready for the next computing revolution?”, Jul 2026
- PostQuantum, “Quantum Machine Learning in 2026: State of the Field”
- The Quantum Insider, “S&P Analysts Report Quantum Computing Arriving Just as Energy Sector Prepares For a Compute-Driven Future,” 7 Apr 2026
- ARPA-E, “U.S. Department of Energy Announces $30 Million to Use Quantum Computing for Groundbreaking Chemistry and Materials Science Simulations”
- ACS Energy Letters, “Harnessing Quantum Computing for Energy Materials: Opportunities and Challenges”
- Quantinuum, “JPMorgan Chase Advances Constrained Quantum Optimization with New 20-Qubit System”
- The Quantum Insider, “Top Global Banks Exploring Quantum Technologies in 2026,” 27 Mar 2026
- McKinsey, “Quantum communication and computing: elevating the banking sector”
- PYMNTS, “Quantum Can’t Do Much Yet, but Banks Can’t Afford to Wait,” 2026
- *New Scientist*, on possible replication issues in quantum computing research, Aug 2026

Fundamental physics

- arXiv:2605.20417, “Quantum Simulation of Gauge Theories for Particle and Nuclear Physics” (review)
- *Nature Physics*, “Simulating two-dimensional lattice gauge theories on a qudit quantum computer”
- *Communications Physics*, “A universal framework for the quantum simulation of Yang–Mills theory”
- *Scientific Reports*, “Digital quantum simulation of cosmological particle creation with IBM quantum computers”
- arXiv:2105.13849, “Quantum Computing for Inflationary, Dark Energy and Dark Matter Cosmology”
- *Astrophysical Journal*, “Toward Cosmological Simulations of Dark Matter on Quantum Computers”
- *Nature*, “Traversable wormhole dynamics on a quantum processor” (Nov 2022); Quanta Magazine, “Physicists Create a Holographic Wormhole Using a Quantum Computer” (Nov 2022) and “Wormhole Experiment Called Into Question” (Mar 2023)
- Caltech, “Physicists observe wormhole dynamics using a quantum computer”; Google Research, “Making a Dual of a Traversable Wormhole with a Quantum Computer”
- *PRX Quantum* 3, 030333, “Entangled Sensor-Networks for Dark-Matter Searches”
- *Physical Review D*, “Enhancing photon-axion conversion probability with squeezed coherent states”

- *npj Quantum Information*, “Quantum-enhanced dark matter detection with in-cavity control: mitigating the Rayleigh curse”
- ORNL, “Quantum sensor research advances dark matter pursuit,” Feb 2026, published in *Physical Review Research*
- Fermilab News, “Yao Lu receives Early Career Award to harness quantum entanglement for dark matter search,” Apr 2026
- DOE National Quantum Information Science Research Centers, “On the hunt for dark matter, with entangled haloscopes”
- Forbes, “Japanese Scientists Using Quantum Computers To Search For Dark Matter,” 28 Mar 2026
- ScienceDaily, “New quantum gravity theory links entropy, dark energy, and life,” 19 Jul 2026
- arXiv:2511.09124, “The Grand Challenge of Quantum Applications”

A note on sourcing. Reporting on quantum computing includes a large volume of low-quality aggregated content. This volume draws on primary sources (peer-reviewed journals, arXiv preprints, company technical blogs and investor filings, and government documents) and on a small set of specialist outlets with editorial track records — The Quantum Insider, Quantum Computing Report, PostQuantum, HPCwire, C&EN, Phys.org and Quanta Magazine. Where only a vendor source exists for a figure, the text says so.

OPX ADVISORY GROUP

Transform. Excel. Outperform.

Kristian Magnuson
Founder, OpX Advisory Group
opxadvisorygroup.com

The Quantum Threshold · First edition, August 2026. Figures current as of 18 August 2026.
Quantum computing is an emerging capability, not a deployed one. The cryptographic exposure described in Chapter 6 is present-tense and belongs to a named owner in your organization. The computational opportunity described in Chapter 5 is an option, and should be sized like one.

www.ingramcontent.com/pod-product-compliance
Lightning Source LLC
LaVergne TN
LVHW070627061126
851857LV00003B/29
* 9 7 9 8 9 9 7 6 8 8 1 1 0 *